LEVEL 1

YOU READ | I READ

Reptiles

Jennifer Szymanski

NATIONAL GEOGRAPHIC

How to Use This Book

Reading together is fun! When older and younger readers share the experience, it opens the door to new learning. As you read together, talk about what you learn.

YOU READ

This side is for a parent, older sibling, or older friend. Before reading each page, take a look at the words and pictures. Talk about what you see. Point out words that might be hard for the younger reader.

I READ

This side is for the younger reader.

As you read, look for the bolded words. Talk about them before you read.

At the end of each chapter, do the activity together.

Contents

CHAPTER 1

Scales and Scutes

mata mata

These animals look very different from each other. But they are both reptiles! Reptiles are alike in some ways. They all have tough, **scaly** skin.

European green lizard

 Scales are part of a reptile's skin. **Scaly** skin helps reptiles stay safe.

YOU READ

Reptiles have a kind of scales called scutes (SKOOTS). Not all scutes are the same. An alligator's scutes are hard and **bony**. They make an alligator's skin look and feel bumpy.

American alligator

Colombian wood turtle

A turtle's shell is made of **bony** scutes, too. But the shell is shiny and smooth.

green anole

YOU READ

Reptiles shed their skin as they get bigger. The old skin loses its **bright** color. Then it becomes loose and falls off. Some reptiles shed skin in patches. Snakes shed their skin in one piece.

The new skin is **bright** and shiny. But it does not feel slimy! It is dry.

Some reptile scales have **special** jobs. The scales at the end of a rattlesnake's tail are hollow. When the snake shakes its tail, the scales make a rattling noise. The noise scares off other animals.

red diamond rattlesnake

All snakes have **special** clear scales over their eyes. They keep dirt and dust out of the snake's eyes.

YOUR TURN!

All reptiles have scales, but not all scales look the same! Can you match each reptile with its scales?

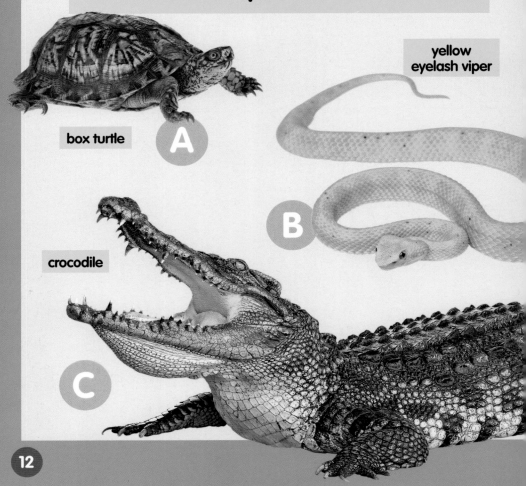

box turtle

A

yellow eyelash viper

B

crocodile

C

CHAPTER 2

On the Move

grass snake skeleton

Reptiles are vertebrates. That means they have bones and a spine inside their bodies. A snake's spine has a lot of small bones. The bones help the snake **slither**. Its whole body touches the ground as it moves.

European adder

European
legless lizard

This animal **slithers** like a
snake. But it is a legless lizard.
Lizards have eyelids and a
thick tongue. Snakes do not!

flying dragon lizard

YOU READ

Reptiles can't fly. But this lizard can **glide** through the air! It has flaps of skin on its sides. The flaps act like wings to help it soar from tree to tree.

Some reptiles are good swimmers. Sea turtles use their flippers to **glide** through the water.

green sea turtle

Other reptiles have legs to help them move quickly on land. A Komodo dragon is a fast runner. It uses its powerful legs to **sprint** short distances as it chases prey.

green basilisk lizard

Some reptiles run on two legs. They **sprint** away from danger. Some can even run across water!

giant tortoise

YOU READ

Giant tortoises are not speedy. Their bodies are heavy! They **crawl** too slowly to run away from predators. Instead, they pull their legs into their shell and hiss to scare off enemies.

Tortoise feet
are wide and flat. Their shape
helps the reptile **crawl** through
grass and mud.

YOUR TURN!

Pretend you are a reptile. How would you move? Would you slither, glide, sprint, or crawl?

CHAPTER 3

Staying Safe

YOU READ

All reptiles have ways to **protect** themselves. A bobtail skink's tail looks a lot like its head. This can confuse other animals. They don't know which end to attack, so they leave the skink alone.

This lizard uses its tail to **protect** itself, too. If something grabs it, the tail falls off. Then the lizard gets away!

lava lizard

YOU READ

Sometimes reptiles stay safe by making themselves seem bigger than they really are. This may be enough to **scare** a predator.

frilled lizard

Some reptiles open their mouths to look bigger. Some can make their bodies bigger. This can **scare** an animal away.

The hognose snake protects itself by playing tricks on other animals. It hisses and sways back and forth. It **pretends** to attack the animal. But it does not bite.

Then the snake rolls over.
It **pretends** to be dead.
The other animal leaves it alone.

29

Some reptiles **fight** to protect themselves and their homes. They use their teeth and claws to keep other animals away.

Komodo dragons

Some reptiles do not **fight**. They hide instead.

timber rattlesnake

YOU READ

Color can help protect a reptile. This coral snake's colorful scales send a warning to other animals that the snake has venom. Venom is poison that could kill another animal.

eastern coral snake

This lizard sits very still in a tree. The shape and **color** of its scales make it look like a leaf.

brown leaf chameleon

YOUR TURN!

Look at the pictures of the reptiles. Tell a story about how each reptile stays safe.

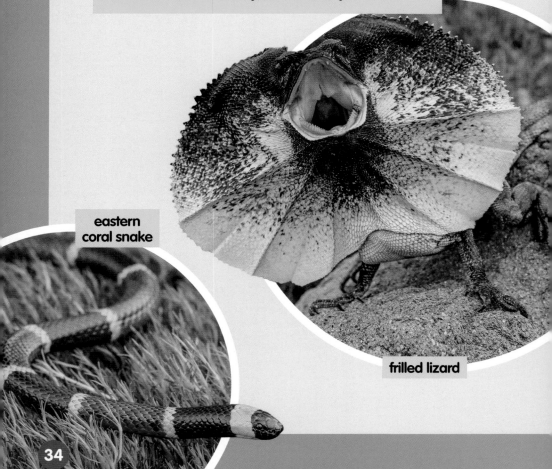

eastern coral snake

frilled lizard

hognose snake

snapping turtle

California
horned lizard

CHAPTER 4

Reptile Homes

YOU
READ

Reptiles can live in lots of different places. They live everywhere except where it is very cold. Many reptiles live in the **desert**.

thorny dragon

Galápagos
land iguana

I READ

There is very little water in the **desert**. Reptiles there get water from the things they eat.

YOU READ Sea turtles live deep in the **ocean**. Some swim long distances to find food. They can stay under the water to eat for many hours at a time.

hawksbill turtle

green
sea turtle

 But reptiles need air. Sea turtles swim to the top of the **ocean** to breathe.

 A reptile's body can't keep itself warm. Its body temperature depends on the temperature of the area around it. This reptile lies in a sunny spot in the **forest**. Its body gets warm.

tuatara

At night, the **forest** cools down. The reptile's body cools down, too.

YOU READ All reptiles need a safe place to have babies. Most reptiles lay eggs. A mother turtle digs a **nest** in the sand. She covers the eggs and leaves them to hatch on their own.

South American river turtle

42

American crocodile

 A mother crocodile stays in the river all day. But at night, she goes to her **nest** to watch her eggs.

43

YOU READ

Reptiles can share a neighborhood with people! A monitor lizard can hunt for fish to eat in a **park** pond.

Red-eared slider turtles
and Chinese three-keeled
pond turtles

Many turtles live in this **park**. People spend the day in the park, too!

YOUR TURN!

Reptiles find places to hide in their homes. The picture shows a swamp. How many reptiles can you find?

American alligators

ANSWER: ten (10)

For April, who loves all creatures scaly and slimy —J.S.

Published by National Geographic Partners, LLC, Washington, DC 20036.

The author and publisher gratefully acknowledge the expert content review of this book by Rebekah Franks, Wildlife Education Director, Heritage Farm Museum and Village, and Thomas K. Pauley, Ph.D., professor emeritus, Marshall University; and the literacy review by Kimberly Gillow, principal, Chelsea School District, Michigan.

Library of Congress Cataloging-in-Publication Data

Names: Szymanski, Jennifer, author.
Title: Reptiles / Jennifer Szymanski.
Description: Washington, D.C. : National
    Geographic Kids, 2022. | Series:
    National geographic readers | Audience:
    Ages 4-6 | Audience: Grades K-1
Identifiers: LCCN 2019055276 (print) | LCCN
    2019055277 (ebook) | ISBN
    9781426338830 (paperback) | ISBN
    9781426338847 (library binding) | ISBN
    9781426338854 (ebook)
Subjects: LCSH: Reptiles--Juvenile literature.
Classification: LCC QL644.2 .S98 2022 (print) |
    LCC QL644.2 (ebook) | DDC 597.9--dc23
LC record available at https://lccn.loc
    .gov/2019055276
LC ebook record available at https://lccn.loc
    .gov/2019055277

Author's Note:
The cover shows a chameleon forest dragon, the title page features a marine iguana, and the contents page shows a box turtle.

Photo Credits
AS: Adobe Stock; GI: Getty Images;
SS: Shutterstock
Cover, lessydoang/GI; top border (throughout), abeadev/SS; 1, ANDREYGUDKOV/GI; 3, Tony Campbell/SS; 4, R. Andrew Odum/GI; 4-5, MF Photo/SS; 6, JG1153/GI; 6 (inset), Cloudia Spinner/AS; 7, Pete Oxford/Minden Pictures; 8, damaloney/GI; 9, Tongho58/GI; 10-11, mgkuijpers/AS; 12 (UP LE), Photo2008a/Dreamstime; 12 (UP RT), Eric Isselée/SS; 12 (LO), dangdumrong/SS; 13 (UP), Mr Preecha/AS; 13 (LO LE), Wirepec/AS; 13 (LO RT), nikonuser66/AS; 14 (UP), argot/AS; 14 (LO), Uryadnikov Sergey/AS; 15, Terry Mathews/Alamy Stock Photo; 16, Satoshi Kuribayashi/Minden Pictures; 17, Steve De Neef/National Geographic Image Collection; 18, Uryadnikov Sergey/AS; 19, Andres Morya/Visuals Unlimited, Inc./GI; 20-21, CraigRJD/GI; 21, PlanetEarthPictures/AS; 22, Fat Camera/GI; 23 (UP), Rebecca Nelson/GI; 23 (CTR), Sergiy Bykhunenko/AS; 23 (LO), rajesh bhand/GI; 24, Stephanie Jackson - Australian wildlife collection/Alamy Stock Photo; 25, City Escapes Nature Photo/SS; 26-27, Ken Griffiths/Alamy Stock Photo; 28, Jay Ondreicka/SS; 29, Kenneth M. Highfill/Science Source; 30-31, Uryadnikov Sergey/AS; 31, Robert Hamilton/Alamy Stock Photo; 32, Liz/AS; 33, Piotr Naskrecki/Minden Pictures; 34 (LE), Mark Kostich/AS; 34 (RT), Matt Cornish/SS; 35 (UP LE), Gerold & Cynthia Merker/GI; 35 (UP RT), Proseuxomai/Dreamstime; 35 (LO), photokh/GI; 36, Michael and Patricia Fogden/Minden Pictures; 37, USO/GI; 38-39, Jag_cz/AS; 39, Colors and Shapes of Underwater World/GI; 40-41, Mark Carwardine/Nature Picture Library; 42, Claus Meyer/Minden Pictures; 43, Helmut Corneli/imageBROKER/AS; 44, carstenbrandt/GI; 45, font83/GI; 46-47, NNehring/GI

Printed in the United States of America
22/WOR/1